Leaping Learners Education, LLC

For more information and resources visit us at:
www.leapinglearnersed.com

Every attempt has been made to credit each photo. Please contact us if there has been an error and we will resolve the issue.

Photo Credits

Monkey graphics © dreamblack46/stock.adobe.com

Cover, © Sergey Uryadnikov/shutterstock.com; Title Page, © Onyx9 shutterstock.com; Pg. 1-2 © Onyx9/shutterstock.com; Pg. 3-4 © max_776/stock.adobe.com; Pg. 5-6 © GUDKOV ANDREY/shutterstock.com; Pg. 7-8 © Kiki Dohmeier/shutterstock.com; Pg. 9-10 © N. F. Photography/shutterstock.com; Pg. 11-12 © dossyl/shutterstock.com; Pg. 13 © Eric Gevaert/shutterstock.com; Pg. 14 © bimserd/shutterstock.com; Pg. 15-16 © Patrick Rolands/ shutterstock.com; Pg. 17 ©Kiki Dohmeie/shutterstock.com; Pg. 18 © Styve Reineck/ shutterstock.com; Pg. 19-20 © Simon Eeman/shutterstock.com; Pg. 21-22 © KiltedArab/ shutterstock.com;
23-24 © Rich Carey/shutterstock.com; Pg. 25 © nikolae/shutterstock.com, © Mari C/ shutterstock.com

*All designs by Sean Bulger*
*All other pictures by Sean Bulger or royalty-free from Pixabay.com*

*ISBN*
*978-1-948569-15-6*

Dear Parents and Guardians,

Thank you for purchasing a *Matt Learns About* series book! After teaching students from kindergarten to second grade for more than seven years, I became frustrated by the lack of engaging books my students could read independently. To help my students engage with nonfiction topics, my wife and I decided to write nonfiction books for children. We hope to inspire young children to learn about the natural world.

Here at Leaping Learners Education, LLC, we have three main goals:

1. Spark young readers' curiosity about the natural world
2. Develop critical independent reading skills at an early age
3. Develop reading comprehension skills before and after reading

We hope your child enjoys learning with Matt. If you or your children are interested in a topic we have not written about yet, send us an email with your topic, and maybe your book will be next!

Thank you,

Sean Bulger, Ed.M

www.leapinglearnersed.com

## Reading Suggestions:

Before reading this book, encourage your children to do a "picture walk," where they skim through the book and look at the pictures to help them think about what they already know about the topic. Thinking about what they already know helps children get excited about learning more facts and begin reading with some confidence.

Preview any new vocabulary words with your child. Key vocabulary words are found on the last few pages of the book. Have your child use the new phrase in their own words to see if they understand the definition.

After previewing the book, encourage your child to read the book independently more than once. After they have read it, ask them specific questions related to the information in the book. Encourage them to go back and reread the relevant section in the book to retrieve the answer in case they forgot the facts.

Finally, see if your child can complete the reading comprehension exercises at the end of the book to strengthen their understanding of the topic!

This book is best for ages 6-8
but. . .
Please be mindful that reading levels are a guide and vary depending on a child's skills and needs.

# Matt Learns About . . . Gorillas

Written by Sean and Anicia Bulger

# Table of Contents

Hi! My name is Matt.
I love to discover and
learn new things. In
this book, we will learn
about gorillas. Let's go!

# Introduction

Have you ever seen a gorilla?

They are rainforest animals.

# Habitat

Gorillas live only in African **rainforests**.

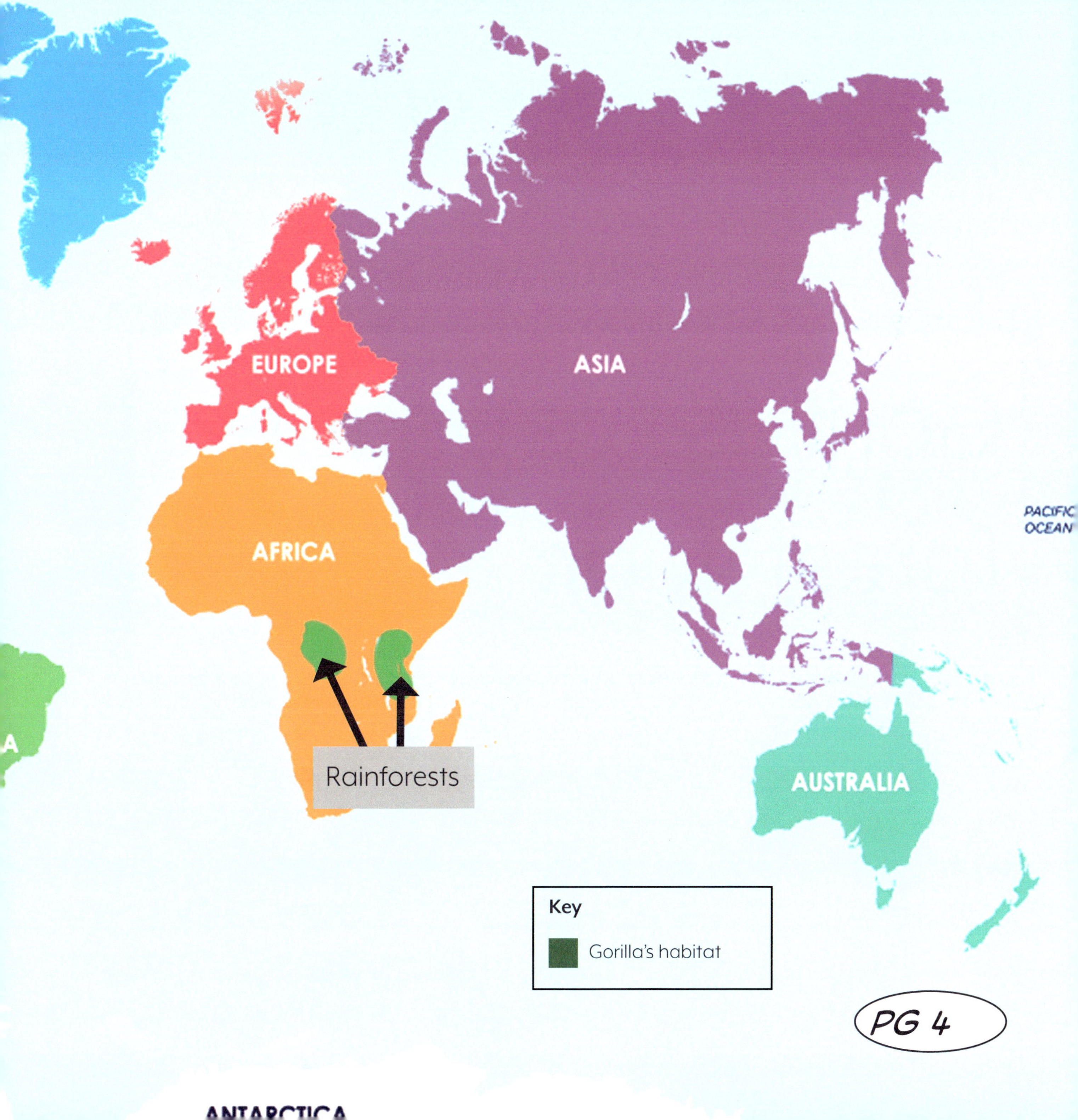
EUROPE
ASIA
AFRICA
A
PACIFIC OCEAN
AUSTRALIA
Rainforests
Key
Gorilla's habitat
ANTARCTICA

Gorillas live on the forest floor.

Gorillas live in groups called **troops**. They sleep on the forest floor. They sleep in a different part of the forest every night.

# Body

Gorillas are big animals and can weigh up to four hundred pounds.

Gorillas use their legs and arms to walk on all fours. Their arms are longer than their legs.

Legs
Arms
PG 12

Gorillas have black fur. Adult male gorillas have gray fur on their back. They are called **silverback** gorillas.

The strongest silverback gorilla is the leader of the troop.

Like people, gorillas have two eyes, two ears, a nose, and a mouth.

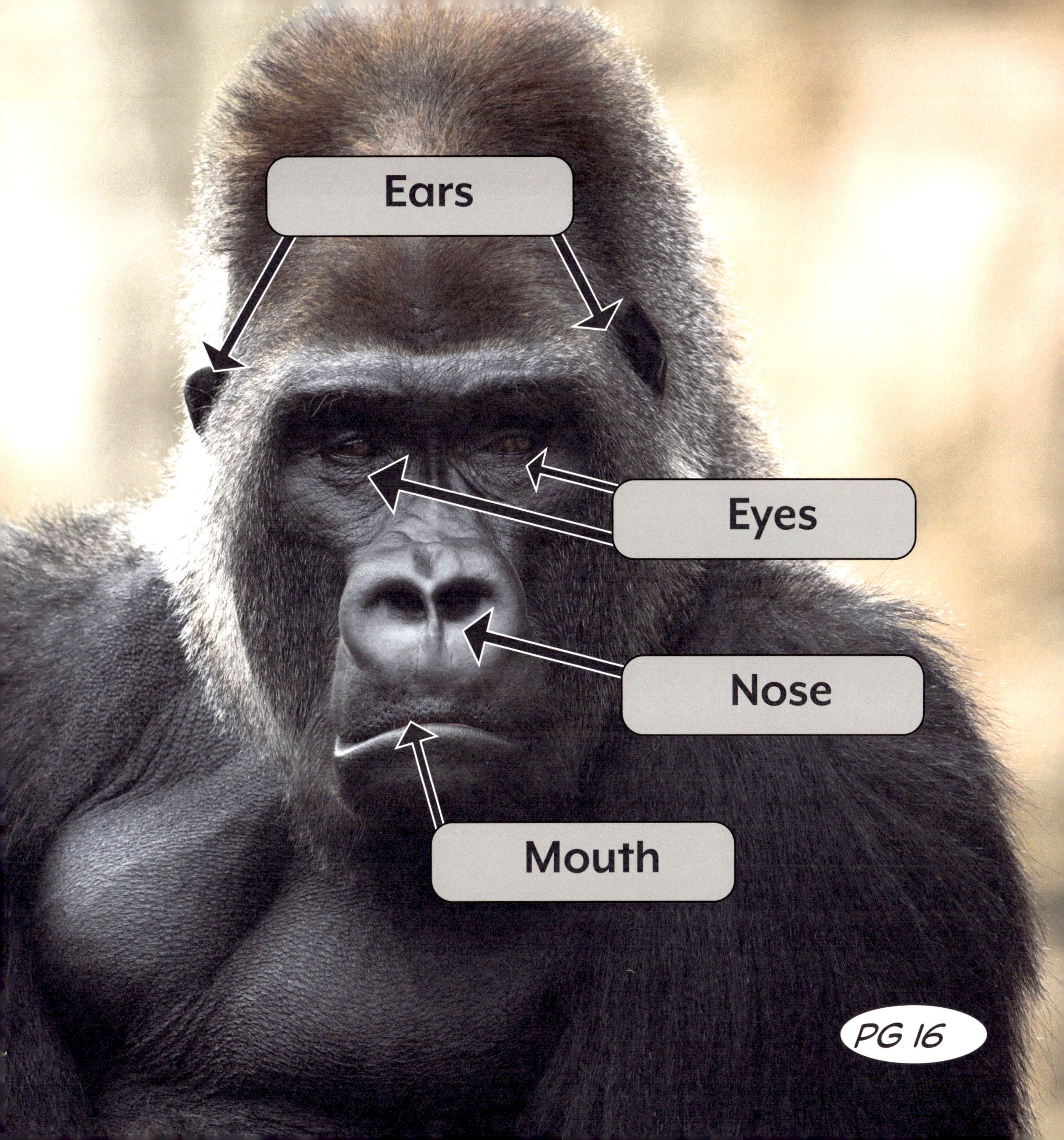
Ears
Eyes
Nose
Mouth

Like people, gorillas have hands. They have five fingers with fingernails!

Gorillas have an **opposable** big toe on each foot and an opposable thumb on each hand. This means they can grab things with their feet and hands.

Tricky word:
opposable

say...
uh-pose-able

# Food

Gorillas eat mostly berries, fruit, flowers, and **bark**.

# Baby Gorillas

Gorillas have one baby every four years. The babies are small and the mother takes care of them for three years.

Baby gorillas like to wrestle and play together.

# Gorillas in Danger

Gorillas are **endangered** because humans are cutting down rainforests. When rainforests are cut down, gorillas have no place to live.

Gorillas use sticks as tools.

Koko the gorilla learned sign language!

The heaviest gorilla weighed 860 pounds!

# Glossary

A glossary tells the reader the meaning of important new words.

**Rainforest** – Tropical forest with many plants and animals

**Troop** – A group of gorillas that live together

**Opposable** – When one finger can touch other fingers and grab things

**Bark** – Hard covering of trees

**Silverback** – Adult male gorilla

**Endangered** – When only a small number of a kind of animal are left on earth

Draw a picture of a gorilla.

Draw a picture of a gorilla eating.

The gorilla is eating a ____________________.

# Quiz

**1. Where do gorillas sleep?**

a. On the forest floor

b. In trees

c. Underground

**2. What color is an adult male gorilla's back?**

a. Silver

b. Brown

c. Tan

**3. What is the main idea of the section called "Gorillas in Danger"?**

a. Why gorillas are in danger

b. How gorillas are dangerous

c. How to catch a gorilla

Common core standards:
RI. 1. 1 - Questions 1, 2
RI. 1. 2 - Question 3

**4. What do gorillas eat?**

a. People

b. Sloths

c. Fruit

**5. Why are gorillas endangered?**

a. They are sick

b. People are cutting down the rainforest

c. They are not endangered

**6. What do the labels on page 12 teach you about?**

a. Body parts of a gorilla

b. How gorillas smile

c. Where gorillas live

Common core standards:
RI. 1.1 - Questions 4, 5
RI. 1.8 - Question 6

# Want to learn about ocean animals? Check out the "Fay Learns About..." Series!

Great for emerging readers ages 6-8

# Want to learn about Farm Animals? Check out the "Katie Teaches you About..." Series!

## Great for early readers ages 4-6

# Want to learn about colors? Check out the "Clayton Teaches you About..." Series!

Great for early readers ages 4-6